# Number Games Around the World

BY MARC GAVE

## Table of Contents

# Introduction

Chess was a popular game of royalty, as this painting from the early 1300s shows.

Do you like to work with numbers? Or do you sigh with relief when you finish your math homework?

Numbers are an essential part of daily life. Just think of all the things you do that involve numbers. Numbers can be fun, too. For thousands of years, people all around the world have been using numbers to play games.

This painting shows the queen Nefertiti, an Egyptian queen from the fourteenth century, playing a game called senet.

At first, number games were very simple. The first game boards were drawn on the ground with sticks. Stones were used as playing pieces. With time and the spread of games to different cultures, the rules sometimes changed. These changes, in turn, often changed the game itself.

**It's a FACT!**

Chess is a famous game in which each piece can be moved only a certain number of squares. In existence since at least 500 A.D., chess is believed to have been first played in India. It is also thought that the pieces represented different parts of the Indian army. Many of the rules of chess have changed over time.

Many people enjoy playing chess today. This boy and his father are an example.

# 1 Magic Squares

Magic squares are similar to crossword puzzles. You can think of them as the crossword puzzles of math.

Magic squares date back thousands of years to ancient Egypt and Greece. They were also played in the Arab world and in China in the twelfth century. In more modern times, magic squares were a favorite of Benjamin Franklin. He loved to make his own.

The idea behind magic squares is simple. No matter how you add a series of numbers—**vertically**, **horizontally**, or **diagonally**—the sum is the same.

**It's a FACT!**

Some people claim that magic squares date from 2200 B.C., when they were played by the Chinese. According to legend, the pattern was first revealed on the shell of a turtle that crawled out of the Lo River.

| | | |
|---|---|---|
| 4 | 9 | 2 |
| 3 | 5 | 7 |
| 8 | 1 | 6 |

In "magic nine," each of the numbers from one to nine must be placed so that the sum of three numbers in any line is 15.

**SOLVE THIS!**

**1 Can you solve these magic squares?**

**A.**

| | | |
|---|---|---|
| 8 | | 4 |
| | | |
| 6 | 7 | 2 |

**B.**

| | | |
|---|---|---|
| 2 | | 6 |
| | 5 | |
| 4 | | 8 |

**C.**

| | | | |
|---|---|---|---|
| 7 | | 1 | 14 |
| 2 | 13 | | |
| 16 | 3 | 10 | 5 |
| | | | 4 |

## CHAPTER 1

# Make Your Own Magic Nine Game

## What you need:

- two square pieces of stiff cardboard (one about 6 by 6 inches and the other about 4 by 4 inches)
- ruler
- quarter or similar-size disk
- pencil
- crayons or markers (optional)
- scissors

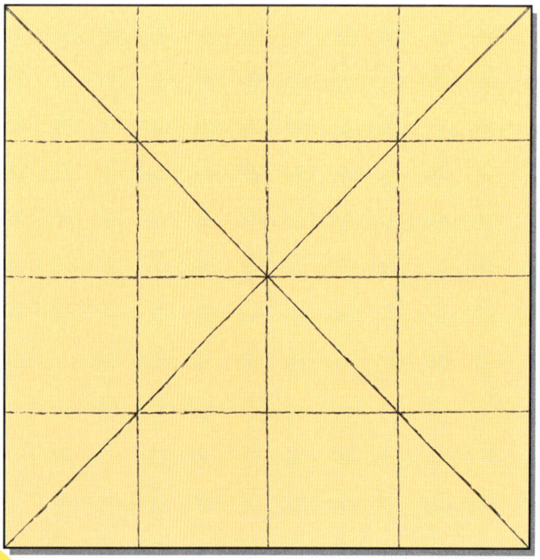

## What to do:

1. Use a ruler to mark up the larger piece of cardboard so that it has four squares across and four squares down. Make your guidelines light unless you want them to form a pattern on your finished game. This is your game board.

2. On the game board, trace around the quarter nine times, as shown.

3. On the smaller piece of cardboard, trace around the quarter nine times and then cut out the nine disks.

4. Color the disks or leave them plain. Write the numbers 1 through 9 on them.

5. Decorate your game board if you wish.

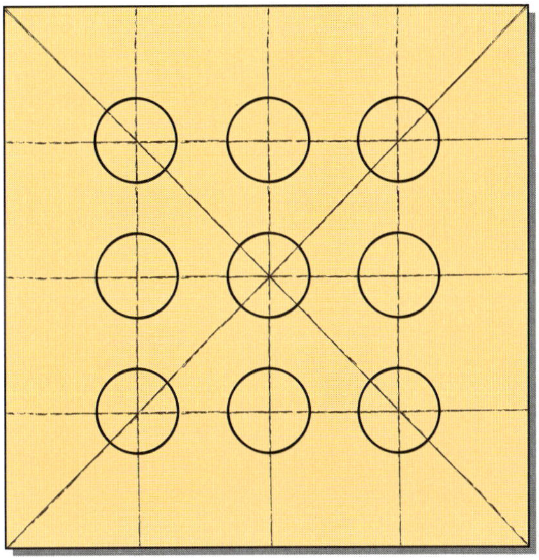

## How to play:

1. Divide the disks between you and another player. One player should take the odd numbers, the other the even numbers. (You can switch off from game to game.)

2. The person with the odd numbers goes first, putting a disk on any of the game-board circles.

3. Players take turns until one player completes a row of three numbers that add up to 15. This player is the winner.

# 2 Tangrams

A tangram is a Chinese game that is played with seven geometric shapes that form a large square. The shapes include two large triangles, one medium triangle, two small triangles, a small square, and a **parallelogram** (pair-uh-LEHL-uh-gram). Each of these shapes is called a tan. The object of the game is to make a **tangram**, or design, using all seven shapes.



square

triangle

parallelogram

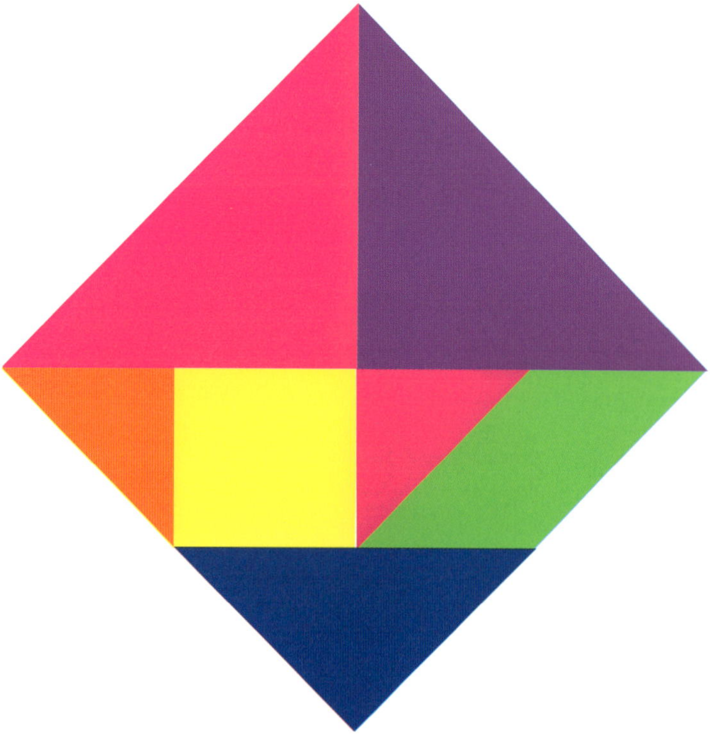

The designs can be a variety of things: animals, people, vehicles, for example. So far, more than 1,600 tangrams have been assembled. You can make your own tangrams by following the designs provided in books available in your library or a bookstore.

**It's a FACT!**

John Quincy Adams (the second president of the United States), Napoleon Bonaparte (ruler of France from 1799 to 1814), Edgar Allan Poe (the mystery writer), and Lewis Carroll (creator of *Alice in Wonderland*) are just a few of the famous people who enjoyed tangrams.

sailboat

person running

person walking

bird in flight

## CHAPTER 2

## Make Your Own Tangram

### What you need:

- piece of poster paper or cardboard about 3½ inches square
- ruler
- pencil
- scissors
- crayons or markers

**SOLVE THIS!**

2 Can you figure out how to put the square back together?

## What to do:

1. Draw a diagonal line across the square. Cut along the line to make two large triangles.

2. Fold one large triangle in half as shown. Cut along the fold.

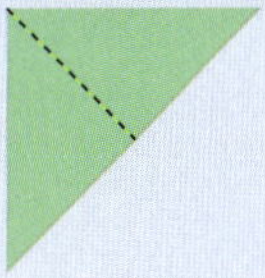

3. Fold the point of the other large triangle as shown. Cut along that fold, too.

4. Fold the larger piece in half and cut it.

5. Fold one of these pieces as shown and cut along the fold.

6. Fold the other small piece as shown and cut along the fold.

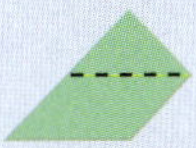

## How to play:

Make your own design. Remember that you must use all seven tans in one picture.

# 3 King Shamba's Game

This is a traditional wari board.

King Shamba's game, and others like it, have been played since the days of ancient Egypt. Games of this kind—known as **mankala** (man-KAH-lah) games—have turned up in Asia and Africa. Also known as **wari** (WAH-ree), the game reached South America and the West Indies when Africans arrived there.

The origin of King Shamba's game is thought to be in a problem the king faced. His people, who lived along the Congo River in Africa, were always arguing. They argued among themselves, as well as with their neighbors. Finally, no one would trade with them. Realizing that he had to do something, King Shamba invented a game that required following rules, communicating, and reaching agreement. His people loved to play it, and as a result, they became peace-loving.

If you can get a traditional wari board, you will be able to play in style. But you can still enjoy the game with your own homemade board.

## Make Your Own Wari Board

### What you need:

- empty carton from a dozen eggs
- markers or paints (optional)
- two paper cups or small bowls
- 48 small game pieces (beads, dried beans, etc.)

### What to do:

1. Cut off the top of the egg carton, unless you want to store your game pieces in it.
2. Decorate the carton if you wish to.

↑ decorated wari board

## How to Play:

In this version of the game, the goal is to capture more pieces (at least 25) than the other player.

1. The board is placed between Player A and Player B so that a row of 6 holes—called houses—faces each player.

2. The players put 4 game pieces in each of the 12 houses.

3. Each player places a cup to the right of his or her six houses to hold captured pieces.

**4.** Player A removes all 4 pieces from any house on his or her side. Moving in a **counterclockwise** direction, he or she drops 1 of the 4 pieces in each of the next 4 houses. Depending on where Player A starts, none, some, or all of the pieces may go into Player B's houses.

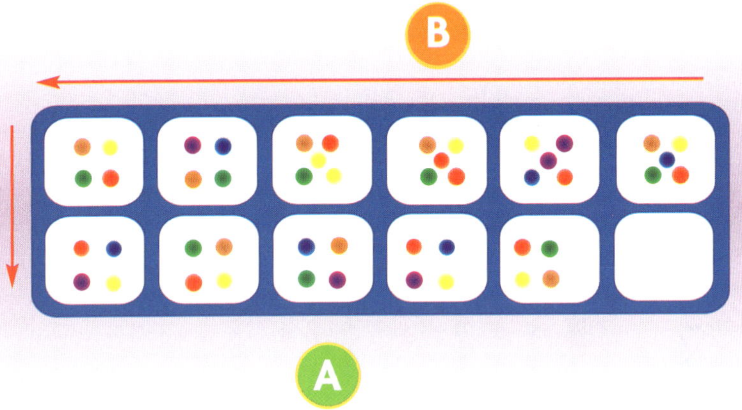


**5.** When it's Player B's turn, he or she takes all of the pieces from any house on his or her side of the board and does the same.

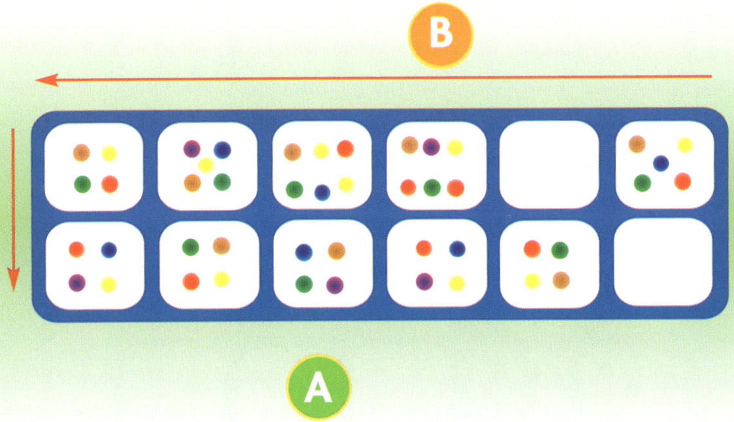

6. If on any turn, Player A drops his or her last piece into a house on Player B's side of the board that contains 1 or 2 pieces—so that it now has 2 or 3 pieces—Player A captures all the pieces in that house.

7. Player A also captures the pieces from any other houses in an unbroken row before the last house that now also has 2 or 3 pieces. Player A can capture pieces from all the houses marked with an arrow.

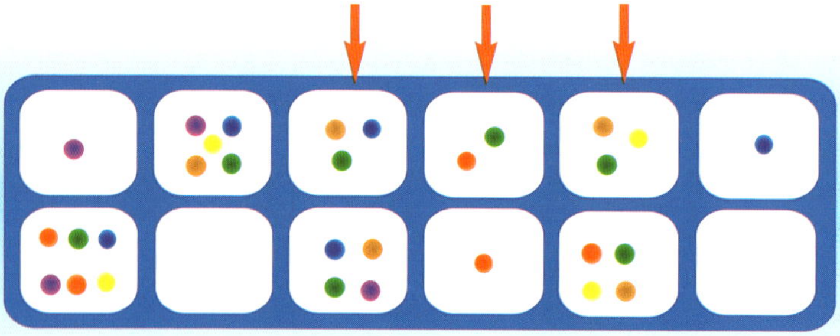

8. If all Player B's houses are empty, Player A must make a move that will give Player B more pieces to play with.

If Player A can't do that, he or she captures all the pieces that remain on his or her side of the board and the game ends.

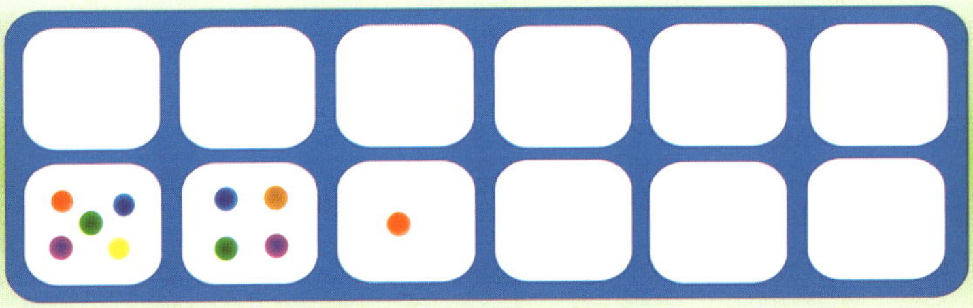

9. A game also may end when there are just a few pieces left on each side of the board and players agree to stop the game. Each player adds the pieces on his or her side of the board to the captured ones. The player with at least 25 pieces wins. If each player has 24 pieces, the game ends in a tie.

Players agree to stop the game.

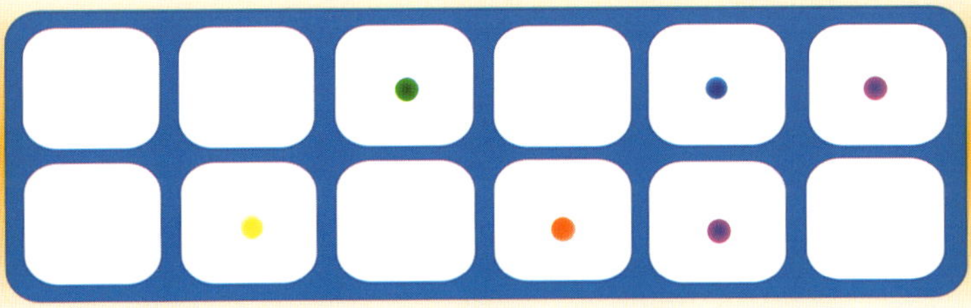

# 4 Secret Codes

The use of **codes** to write messages has a long history. The ancient Egyptians used a system of symbols to represent objects and numbers. This picture script, called hieroglyphics (hye-ruh-GLIH-fiks), is believed to be the oldest form of writing.

The hieroglyphics shown here represent numbers. They were used by ancient Egyptians to count such things as people, cattle, or farm crops.

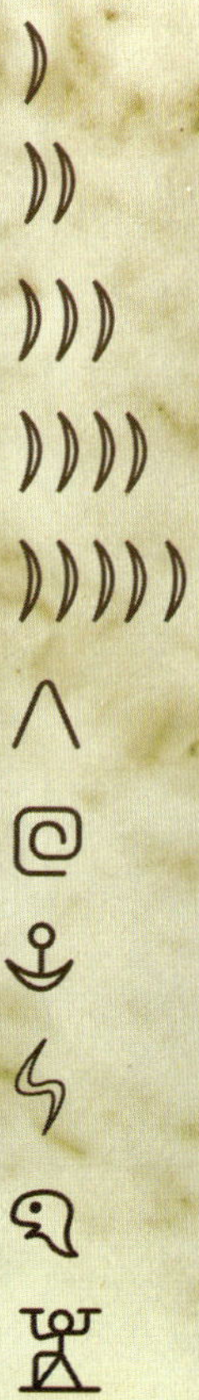

| | |
|---|---|
| 1 | (a papyrus leaf) |
| 2 | |
| 3 | |
| 4 | |
| 5 | |
| 10 | (a bent leaf) |
| 100 | (a curled rope) |
| 1,000 | (a lotus flower) |
| 10,000 | (a snake) |
| 100,000 | (a tadpole) |
| 1,000,000 | (a surprised scribe) |

The meaning of hieroglyphics remained a mystery until long after the fall of ancient Egyptian civilization. Then in 1799, a stone inscribed with a message written in hieroglyphics and two other languages was discovered in Egypt. One of the languages was ancient Greek. Translators used the ancient Greek to decode the hieroglyphics. The stone was named the Rosetta Stone by its discoverers.

Many secret codes use letters of the alphabet instead of objects to represent numbers. You can create your own code using this idea or its opposite: using numbers to represent letters of the alphabet.

One of the simplest codes assigns a number to each letter of the alphabet, starting with A = 1 and ending with Z = 26. To make this code harder to figure out, you can add or subtract a set number from the numbers 1 through 26. For example, you can add 5 to each, making A = 6 and Z = 31.

| | CODE 1 | CODE 2 |
|---|---|---|
| A | 1 | 6 |
| B | 2 | 7 |
| C | 3 | 8 |
| D | 4 | 9 |
| E | 5 | 10 |
| F | 6 | 11 |
| G | 7 | 12 |
| H | 8 | 13 |
| I | 9 | 14 |
| J | 10 | 15 |
| K | 11 | 16 |
| L | 12 | 17 |
| M | 13 | 18 |
| N | 14 | 19 |
| O | 15 | 20 |
| P | 16 | 21 |
| Q | 17 | 22 |
| R | 18 | 23 |
| S | 19 | 24 |
| T | 20 | 25 |
| U | 21 | 26 |
| V | 22 | 27 |
| W | 23 | 28 |
| X | 24 | 29 |
| Y | 25 | 30 |
| Z | 26 | 31 |

You can also assign the numbers backward, so that Z = 1 and A = 26. To make that more difficult, you can add or subtract another number. For example, if Z = 1 and you add 4, Z becomes 5. U becomes 10. Can you write this code? How about making your own code using numbers for letters? See if your friends or family members can decode your secret messages. If they can, you will be able to write secret messages to each other.

| | CODE 3 | CODE 4 | YOUR CODE |
|---|---|---|---|
| A | 26 | | |
| B | 25 | | |
| C | 24 | | |
| D | 23 | | |
| E | 22 | | |
| F | 21 | | |
| G | 20 | | |
| H | 19 | | |
| I | 18 | | |
| J | 17 | | |
| K | 16 | | |
| L | 15 | | |
| M | 14 | | |
| N | 13 | | |
| O | 12 | | |
| P | 11 | | |
| Q | 10 | | |
| R | 9 | | |
| S | 8 | | |
| T | 7 | | |
| U | 6 | 10 | |
| V | 5 | | |
| W | 4 | | |
| X | 3 | | |
| Y | 2 | | |
| Z | 1 | 5 | |

## SOLVE THIS!

3 Now that you know how to use numbers to create secret messages, you can do it the other way around. You can assign each number a letter and then solve the problems below.

| | | |
|---|---|---|
| 1 = A | 5 = E | 8 = H |
| 2 = B | 6 = F | 9 = I |
| 3 = C | 7 = G | 10 = J |
| 4 = D | | |

1. B ÷ H = ?
2. G x D = ?
3. F + E + H = ?
4. (E x I) - (B x A) = ?
5. (F x A) x (I x H) = ?
6. (C x J) ÷ (J x A) = ?

Now try to make your own letter codes.

# 5 Fibonacci Series

Leonardo Fibonacci (lay-oh-NAR-doh fih-boh-NAH-chee) was born in Pisa, the Italian city famous for its leaning tower, in 1170. His father was an official in North Africa. While he lived there, young Fibonacci learned to write the **Arabic numerals** we use today. Up until then, he had used **Roman numerals**.

When Fibonacci returned to Italy, he continued to pursue his interest in math by writing books on the subject, using the new numerals he had learned.

Fibonacci was also interested in nature, and it wasn't long before he found a way to combine this interest with his knowledge of math. He noticed that there was a certain **sequence**, or pattern, of numbers that appeared often in nature.

The lily has 3 petals.

The columbine has 5 petals.

The ragwort has 13 petals. Count them!

## CHAPTER 5

The sequence is:

1, 1, 2, 3, 5, 8, 13, 21, 34 . . .

Can you see the key to the pattern? To understand the sequence, start with 1. To get the next number in the series, add the current number and the number that precedes it. Because the number that precedes 1 is 0 and 1 + 0 = 1, there are two 1s in the series.

Continuing this process gives the following:

1 + 1 = 2

2 + 1 = 3

3 + 2 = 5

5 + 3 = 8

8 + 5 = 13

13 + 8 = 21

21 + 13 = 34

**SOLVE THIS!**

**4** **What would the next three numbers in the Fibonacci series be?**

**5** **Fibonacci numbers also apply to the human body, specifically the hands. Can you identify in what way?**

# 6 Counting with Knots

This elaborate quipu is typical of the way the Inca kept number records and sent messages. Quipus had to be made up and decoded by specially trained officers. The method of decoding is not completely understood today.

Can you imagine counting things with knots? That's exactly what the Inca people did. The Inca empire began in southern Peru. It grew to include much of northwest South America by the time the Spanish conquered the area in 1530.

The Inca kept records using a system of knots called **quipu** (KEE-poo). They tied the knots in lengths of string made of cotton or wool. The knots helped the Inca keep track of harvest information, taxes, population, the days until an event—almost anything that could be counted.

## CHAPTER 6

A typical quipu has several cords. The color of each cord, its placement, its size, and the knots in it were all significant to the record or the message the quipu contained.

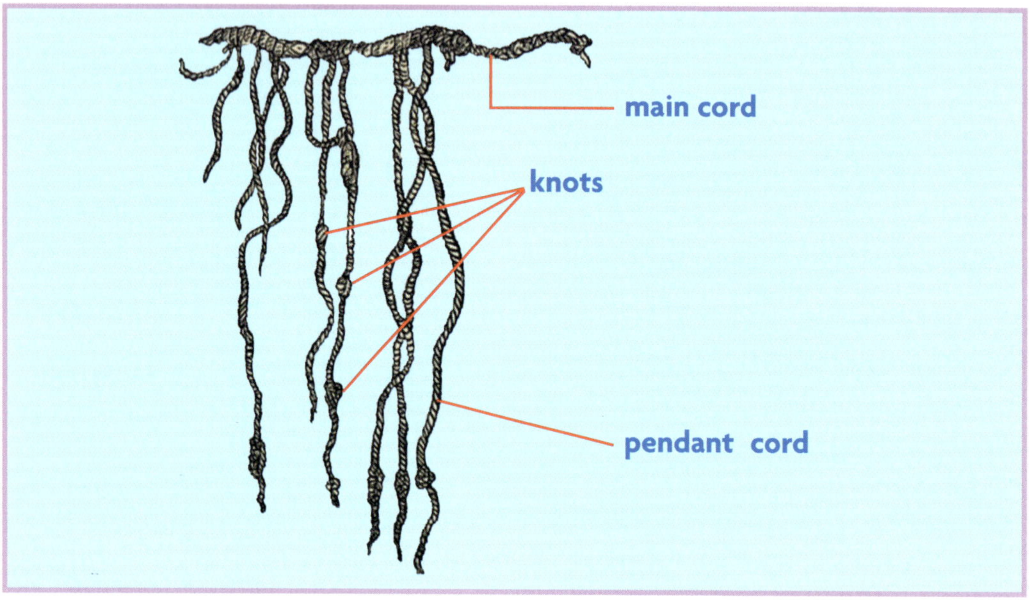


## 1. Main cord

The main cord is longer and thicker than all the other cords. At one end is a loop, and at the other end is a knot. The main cord is held horizontally (side to side). All the other cords are attached to it.

## 2. Pendant cords

The word "pendant" means "hanging." Pendant cords hang down from the main cord. The unlooped end of a pendant cord is passed around the main cord, through its own loop, and pulled tight.

### 3. Knots

Knots are tied in the pendant cords. The Inca seem to have used a counting system in base 10 like ours. There were different levels in each pendant cord, depending on how many numbers needed to be included. The lowest level was the units; the next-higher level was tens. Higher than tens was hundreds; higher than hundreds was thousands; and so on. The knots at each level showed the number in the place.

### 4. Other cords

Sometimes a dangle cord was attached to the loop end of the main cord. It could have pendant cords of its own. The quipu keeper would record other information, such as the number of days of rain in the year, on it.

A top cord was sometimes tied around the loops of several pendant cords. It was used for adding up the information provided by the knots on a group of pendant cords.

**SOLVE THIS!**

**6** **Numberville has 304 people. Mathtown has 296. Digitopolis has 751. Using the information about the cords and knots on a quipu, draw a diagram of the quipu you would make to record these numbers.**

# 7 Conclusion

You have learned that games involving numbers have been played for thousands of years all over the world. The games vary in numbers of players, materials, and complexity.

Many number games have their origins in ancient civilizations. Other games come from the imaginations of people who like to work with numbers and know that numbers can be entertaining. Today we use numbers to play all sorts of games. Have you ever played Monopoly? Parcheesi? Dominoes? If so, you were carrying on a long-standing tradition—having fun with numbers.

A boy and a school volunteer play a math game together.

A family plays a card game together.

Dominoes is a game that is played all over the world.

Europe

Middle East

North America

# SOLVE THIS! Answers

1 Page 5

| | | |
|---|---|---|
| 8 | 3 | 4 |
| 1 | 5 | 9 |
| 6 | 7 | 2 |

| | | |
|---|---|---|
| 2 | 7 | 6 |
| 9 | 5 | 1 |
| 4 | 3 | 8 |

| | | | |
|---|---|---|---|
| 7 | 12 | 1 | 14 |
| 2 | 13 | 8 | 11 |
| 16 | 3 | 10 | 5 |
| 9 | 6 | 15 | 4 |

2 Page 10

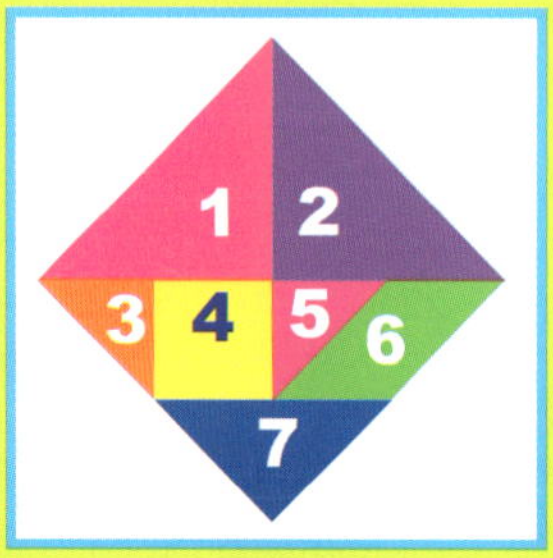


3 Page 22

**1/4; 28; 19; 43; 432; 3**

4 Page 24

**55; 89; 144**

5 Page 24

**2 hands, 5 fingers,**
**3 parts to each finger,**
**2 knuckles**

6 Page 27

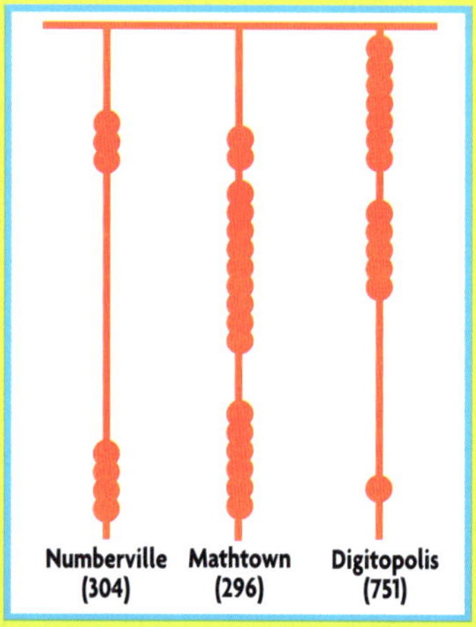

# Glossary

| | |
|---|---|
| Arabic numerals | our familiar way of writing numbers—1, 2, 3, etc.—which was developed in the Arab world |
| code | secret writing that uses some kind of symbol for a letter or a word |
| counterclockwise | moving in a circle in the opposite direction from the hands of a clock |
| diagonally | on a slant |
| horizontally | across |
| mankala | any of a number of ancient games played with holes and stones or similar objects |
| parallelogram | shape with two opposite horizontal sides of equal length and two opposite diagonal sides of equal length |
| quipu | system of record keeping using knots developed by the Inca in South America |
| Roman numerals | way of writing numbers developed by the Romans, using I for 1, V for 5, X for 10, and so on |
| sequence | series or pattern, as of numbers: 3, 6, 9, 12 is a sequence in which each number is three greater than the number before |
| tangram | Chinese game that is played with seven geometric shapes that form a large square |
| vertically | up and down |
| wari | type of mankala game from Africa |

# Index